# Mathematical Notation Reference for Computer Science

Daniel Szelogowski

2023

# Preface

For errata, visit **http://danielszelogowski.com/research.php#mathnotation**. If you find any errors, please use the contact form on my homepage or email **dan@zerodevelopment.net**, and I will add them to the errata.

If you would like a digital copy, please email me and provide a proof of purchase; I will send it to you free of charge.

Copyright © 2023 by Daniel Szelogowski

All rights reserved. This book or any portion thereof may not be reproduced or used in any manner whatsoever without the publisher's express permission except for the use of brief quotations in a book review or scholarly journal.

First Printing: 2023
ISBN 978-1-304-80795-3

Lulu Publishing
Janesville, Wisconsin 53545

www.danielszelogowski.com

# Contents

**1 Numbers**   1
    1.1 Basic Operations . . . . . . . . . . . . . . . . . . . . . . . . . . . . . . . . . . . . . . . 1
    1.2 Common Notations . . . . . . . . . . . . . . . . . . . . . . . . . . . . . . . . . . . . . 2
    1.3 Relationships . . . . . . . . . . . . . . . . . . . . . . . . . . . . . . . . . . . . . . . . 2
    1.4 Intervals (Domain and Range) . . . . . . . . . . . . . . . . . . . . . . . . . . . . . . . 2
    1.5 Real Number Set and Subsets . . . . . . . . . . . . . . . . . . . . . . . . . . . . . . . 3
    1.6 Well-Known Real Numbers . . . . . . . . . . . . . . . . . . . . . . . . . . . . . . . . 3
    1.7 Infinity and Beyond . . . . . . . . . . . . . . . . . . . . . . . . . . . . . . . . . . . . 3
    1.8 Other Number Systems . . . . . . . . . . . . . . . . . . . . . . . . . . . . . . . . . . . 3
    1.9 Complex Numbers . . . . . . . . . . . . . . . . . . . . . . . . . . . . . . . . . . . . . 3

**2 Functions**   4
    2.1 Fundamentals . . . . . . . . . . . . . . . . . . . . . . . . . . . . . . . . . . . . . . . . 4
    2.2 Standard Functions . . . . . . . . . . . . . . . . . . . . . . . . . . . . . . . . . . . . . 5
    2.3 Combinatorial Functions . . . . . . . . . . . . . . . . . . . . . . . . . . . . . . . . . . 6
    2.4 Bitwise Functions . . . . . . . . . . . . . . . . . . . . . . . . . . . . . . . . . . . . . . 7
    2.5 Miscellaneous . . . . . . . . . . . . . . . . . . . . . . . . . . . . . . . . . . . . . . . . 7

**3 Logic**   8
    3.1 Boolean Operations and Proof Symbols . . . . . . . . . . . . . . . . . . . . . . . . . . 8
    3.2 Quantifiers . . . . . . . . . . . . . . . . . . . . . . . . . . . . . . . . . . . . . . . . . 8

**4 Collections**   9
    4.1 Sets . . . . . . . . . . . . . . . . . . . . . . . . . . . . . . . . . . . . . . . . . . . . . 9
    4.2 Lists . . . . . . . . . . . . . . . . . . . . . . . . . . . . . . . . . . . . . . . . . . . . . 10
    4.3 Big Sums, Products, etc. . . . . . . . . . . . . . . . . . . . . . . . . . . . . . . . . . . 10
    4.4 Graphs and Trees . . . . . . . . . . . . . . . . . . . . . . . . . . . . . . . . . . . . . . 11

**5 Geometry and Linear Algebra**   12
    5.1 Vectors . . . . . . . . . . . . . . . . . . . . . . . . . . . . . . . . . . . . . . . . . . . 12
    5.2 Matrices . . . . . . . . . . . . . . . . . . . . . . . . . . . . . . . . . . . . . . . . . . . 13
    5.3 Matrix Norms . . . . . . . . . . . . . . . . . . . . . . . . . . . . . . . . . . . . . . . . 14
    5.4 Tensors . . . . . . . . . . . . . . . . . . . . . . . . . . . . . . . . . . . . . . . . . . . 14
    5.5 Groups . . . . . . . . . . . . . . . . . . . . . . . . . . . . . . . . . . . . . . . . . . . . 14
    5.6 Coordinates . . . . . . . . . . . . . . . . . . . . . . . . . . . . . . . . . . . . . . . . . 14
    5.7 Geometry . . . . . . . . . . . . . . . . . . . . . . . . . . . . . . . . . . . . . . . . . . 15

# 6 Calculus — 16
## 6.1 Limits — 16
## 6.2 Derivatives — 16
### 6.2.1 Single Independent Variable, Scalar- or Vector-Valued — 16
### 6.2.2 Multiple Independent Variables, Scalar-Valued — 16
### 6.2.3 Multiple Independent Variables, Vector-Valued — 17
## 6.3 Integration — 17
## 6.4 Convolution and Transforms — 17

# 7 Probability and Statistics — 18
## 7.1 Probability — 18
### 7.1.1 Distributions — 19
## 7.2 Statistics — 20

# 8 Approximation — 21
## 8.1 Asymptotic Relations — 21
## 8.2 Big-O Notation and Relatives — 21

# 9 Letters — 22
## 9.1 Alphabet Decorators — 22
## 9.2 Latin Alphabet (English) — 22
## 9.3 Greek Alphabet — 23

# 10 Academic Language — 24
## 10.1 Latin Abbreviations — 24
## 10.2 Latin Phrases — 25

# About

This text contains a comprehensive reference for many commonly seen mathematical notations used frequently in Computer Science, including symbols (or other relevant notation), names, meanings, and programming equivalents (in Python, if any), alongside the most frequently used alphabets (i.e., Latin and Greek) and Latin abbreviations and phrases commonly used in academic literature. The document was typeset using the LaTeX markup language (similar to Markdown and HTML), one of the most commonly used document preparation systems across the majority of scientific fields.

Additional Resources:

- Mathematical Notation: A Guide for Engineers and Scientists (Edward R. Sheinerman, 2011)
- https://github.com/Experience-Monks/math-as-code/blob/master/PYTHON-README.md
- https://www.math.ucdavis.edu/~anne/WQ2007/mat67-Common_Math_Symbols.pdf
- https://www.overleaf.com/learn/latex/List_of_Greek_letters_and_math_symbols
- https://www.rapidtables.com/math/symbols/Basic_Math_Symbols.html
- https://en.wikipedia.org/wiki/List_of_mathematical_symbols_by_subject
- https://wikieducator.org/Help:LaTeX_Symbol_Tables_-_Mathematics
- https://en.wikipedia.org/wiki/Glossary_of_mathematical_symbols
- https://en.wikipedia.org/wiki/List_of_Latin_phrases

---

# 1 Numbers

## 1.1 Basic Operations

| Symbol | Name | Meaning | Programming Equivalent |
|---|---|---|---|
| $x + y$ | Addition | Sum of two numbers | `x + y` |
| $x - y$ | Subtraction | Difference of two numbers | `x - y` |
| $x \times y$ or $x * y$ or $x \cdot y$ | Multiplication | Product of two numbers | `x * y` |
| $x \div y$ or $x/y$ or $\frac{x}{y}$ | Division/Percentage (%) | Quotient of two numbers | `x / y` |
| $x \bmod y$ | Modulus | Remainder from long division of two numbers | `x % y` (or, to get the quotient and remainder) `q, r = divmod(x, y)` |
| $x\|y$ | Divisibility ($x$ divides $y$) | $y/x$ leaves no remainder | `y % x == 0` |
| $x^n$ | Exponentiation | $x$ raised to the power of $n$ | `x**n` |
| $x : y$ | Ratio | Relationship between two quantities | ```def ratio(x, y):```<br>```    div = math.gcd(x, y)```<br>```    return f"{x/div}:{y/div}"``` |

## 1.2 Common Notations

| Symbol | Name | Meaning | Programming Equivalent |
|---|---|---|---|
| $\pm x$ | Plus-Minus | Indicates a range of values, both positive and negative | A tuple of both variations or range, e.g., (x, -x) |
| $\mp x$ | Minus-Plus | Indicates a range of values, both negative and positive | (-x, x) |
| $+x$ | Unary Plus | The positive value of x | +x |
| $-x$ | Negation (Unary Minus) | The negative value of x | -x |
| $1.23 \times 10^4$ (same as 12300) or $1.23 \times 10^{-4}$ (same as 0.000123) | Scientific Notation | A convenient method of writing very large or small numbers | 1.23e4 (or) 1.23e-4 |
| $0.\overline{3}$ | Repeating Decimal | A decimal fraction where a figure or group of figures is repeated indefinitely | 1/3 (fraction approximation) |

## 1.3 Relationships

| Symbol | Name | Meaning | Programming Equivalent |
|---|---|---|---|
| $x = y$ | Equals | A value is the same as a second value | x = y |
| $x \approx y$ | Approximately Equals | A value is nearly the same as a second value | math.isclose(x, y, rel_tol=1e-9) (or, for $n$ significant digit closeness) math.isclose(x, y, abs_tol=10**-n) |
| $x \pm y$ | Approximate Value | An approximate value ($x$) with some difference threshold ($y$) | x_pm_y = (x+y, x-y) (a tuple indicating the range of values) |
| $x \equiv y$ | Equivalence | A value is equivalent (identical) to a second value | x == y |
| $x \neq y$ | Not Equals | A value is not equivalent to a second value second value | x != y |
| $x > y$ | Greater Than | Same as name | x > y |
| $x < y$ | Less Than | Same as name | x < y |
| $x \geq y$ | Greater Than or Equal To | Same as name | x >= y |
| $x \leq y$ | Less Than or Equal To | Same as name | x <= y |
| $x \gg y$ | Much Greater Than | Significantly greater than by some factor | `def much_greater_than(x, y, factor=2):`<br>`    return x > factor * y` |
| $x \ll y$ | Much Less Than | Significantly less than by some factor | `def much_less_than(x, y, factor=2):`<br>`    return x * factor < y` |
| $x \propto y$ | Proportional To | Same as name | `def are_proportional(set1, set2):`<br>`    s1 = numpy.array(set1)`<br>`    s2 = numpy.array(set2)`<br>`    if s1.size != s2.size or s1.size < 2:`<br>`        return False`<br>`    ratios = s1 / s2`<br>`    return numpy.allclose(ratios, ratios[0])` |

## 1.4 Intervals (Domain and Range)

| Symbol | Meaning | Programming Equivalent |
|---|---|---|
| $a \leq x < b$ | Inequality Expression | a <= x and x < b  (or)  a <= x < b |
| $[a, b]$ | Interval including $a$ and $b$ ($a \leq x \leq b$) | range(a, b+1) |
| $(a, b)$ or $]a, b[$ | Interval excluding $a$ and $b$ ($a < x < b$) | range(a+1, b) |
| $[a, b)$ or $[a, b[$ | Interval including $a$ but excluding $b$ ($a \leq x < b$) | range(a, b) |
| $(a, b]$ or $]a, b]$ | Interval excluding $a$ but including $b$ ($a < x \leq b$) | range(a+1, b+1) |

## 1.5 Real Number Set and Subsets

| Symbol | Name | Meaning | Programming Equivalent |
|---|---|---|---|
| $\mathbb{R}$ | Real Numbers | All rational and irrational numbers | Floating-point numbers in Python |
| $\mathbb{Q}$ | Rational Numbers | All numbers that can be expressed as a fraction | Fractions in Python with `fractions.Fraction` |
| $\mathbb{Z}$ | Integers | Whole numbers, positive, negative, and zero (sometimes $\mathbb{Z}^*$ for positive integers) | Integers in Python |
| $\mathbb{N}$ or $\mathbb{N}_0$ | Natural Numbers | Positive integers (sometimes with zero) | Positive range in Python |

## 1.6 Well-Known Real Numbers

| Symbol | Name | Meaning | Programming Equivalent |
|---|---|---|---|
| $\pi$ | Pi | Ratio of a circle's circumference to its diameter | `math.pi` |
| $\tau$ | Tau | $2\pi$ | `math.tau` |
| $e$ | Euler's Number | Base of the natural logarithm | `math.e` |
| $\gamma$ | Euler-Mascheroni Constant | The limiting difference between the harmonic series and the natural log | `sympy.EulerGamma.evalf()` |
| $\phi$ | Golden Ratio | (1 + sqrt(5)) / 2 | `(1 + math.sqrt(5)) / 2` |

## 1.7 Infinity and Beyond

| Symbol | Name | Meaning | Programming Equivalent |
|---|---|---|---|
| $\infty$ | Infinity | A value greater than any finite number | `math.inf` |
| $-\infty$ | Negative Infinity | A value less than any finite number | `-math.inf` |
| $\omega$ | Ordinal Number | A value representing a rank (1st, 2nd, ...) | N/A |
| $\aleph_0$ | Aleph Null | The (infinite) cardinality of the set of natural numbers ($\aleph_0 = |\mathbb{Z}|$) | N/A |
| $\aleph_1$ | Aleph One | The cardinality of the countable ordinal numbers set ($\omega_1$ or $\Omega$) | N/A |

## 1.8 Other Number Systems

| Symbol | Name | Meaning | Programming Equivalent |
|---|---|---|---|
| $\mathbb{B}$ | Binary | Base-2 (0, 1) | `bin(x)` |
| $\mathbb{O}$ | Octal | Base-8 (0-7) | `oct(x)` |
| $\mathbb{H}$ | Hexadecimal | Base-16 (0-9, A-F) | `hex(x)` |
| $\mathbb{F}_n$ or $GF(n)$ | Finite (Galois) Field | Algebraic field with $n$ elements | `galois.GF(n)` |

## 1.9 Complex Numbers

| Symbol | Name | Meaning | Programming Equivalent |
|---|---|---|---|
| $\mathbb{C}$ | Complex Numbers | Set of all complex numbers | Complex numbers in Python |
| $\mathbb{I}$ | Imaginary Numbers | Set of all imaginary numbers | Imaginary numbers in Python |
| $\mathbb{H}$ | Quaternions | Set of all quaternions | Quaternions in Python |
| $i$ or $\mathbf{i}$ (or $j$) | Imaginary Unit | Square root of -1 | `1j` |
| $ai$ | Imaginary Number | Number with imaginary part | `aj` |
| $a + bi$ or $re^{i\theta}$ (polar form) | Complex Number | Number with real and imaginary parts | `z = complex(a, b)` (or) `z = (a + bj)` |
| $a + bi + cj + dk$ | Quaternion | Number with one real part and three basis vectors/elements | `numpy.quaternion(1, 2, 3, 4)` |
| $\bar{z}$ or $z^*$ (e.g., $a - bi$) | Complex Conjugate | A complex number with the sign of the imaginary part reversed | `complex(a, b).conjugate()` |
| $\Re(a + bi)$ or $Re(z)$ | Real Operator | Returns the Real part ($a$) of a complex number | `complex(a, b).real` |
| $\Im(a + bi)$ or $Im(z)$ | Imaginary Operator | Returns the Imaginary part ($b$) of a complex number | `complex(a, b).imag` |

# 2 Functions

## 2.1 Fundamentals

| Symbol | Name | Meaning | Programming Equivalent |
|---|---|---|---|
| $f(x)$ or $f: x \mapsto y$ or $f: A \to B$ | Function | Maps input (e.g., $x$) to an output ($f(x)$ or a variable; i.e., $x \mapsto f(x)$) | Function definition; e.g., $f(x,y,z) = x^y + z$ is:<br>```def f(x, y, z):```<br>```    return x**y + z```<br>```output = f(1, 2, 3)```<br>(or)<br>```f = lambda x, y, z: x**y + z``` |
| $f(x) = \begin{cases} 1, \text{ if } x \geq 0 \\ 0, \text{ if } x < 0 \end{cases}$ | Piecewise Function | Function with conditional outputs | ```def f(x):```<br>```    return 1 if x >= 0 else 0``` |
| $f^{-1}(x)$ | Function Inverse | The "undo" of a given function | ```f_inv = pynverse.inversefunc(f)```<br>```y = f_inv(x)``` |
| $f(\cdot)$ | Function with an Expected Argument | Function with an unspecified (or placeholder) argument | ```def f(x):```<br>```    return x**2``` |
| $(f \circ g)(x)$ | Function Composition | Composition of functions $f$ and $g$ (i.e., $f(g(x))$) | ```def f(x): return x**2```<br>```def g(x): return x + 1```<br>```def fog(x): return f(g(x))``` |
| $f(x, y, \ldots)$ | Variadic Function | Function with a variable number of arguments | ```def f(*args):```<br>```    return sum(args)``` |
| $id(x)$ | Identity Function | Function that returns its input | ```def id(x):```<br>```    return x``` |
| $f(A)$ | Function Image | Set of all images of elements of set $A$ under $f$ | ```def f(x):```<br>```    return x**2```<br>```A = {1, 2, 3}```<br>```image = {f(x) for x in A}``` |
| $f^{(n)}(x)$ | Iterated Function (i.e., self-composed) | Function iterated $n$ times | ```def iterate(f, n):```<br>```    return lambda x: \```<br>```    f(iterate(f, n-1)(x)) \```<br>```    if n > 0 else x```<br>```fn = iterate(f=lambda x: x*2, n=3)``` |
| $\chi_A(x)$ | Characteristic Function | Function that indicates membership of $x$ in set $A$ | ```def chi_A(x, A):```<br>```    return 1 if x in A else 0``` |
| $p(x)$ | Polynomial | Polynomial function of $x$ | ```def p(x):```<br>```    return a*x**2 + b*x + c``` |
| $\lambda x : f(x)$ | Lambda Function | An anonymous function defined by an expression | ```lambda x: x**2``` |
| $f \rightarrowtail g$ or $f \hookrightarrow g$ | Injective (One-to-One) Function | Every element of the function's codomain is mapped by at most one element of its domain | ```def is_injective(func, domain):```<br>```    images = set(func(x) for x```<br>```        in domain)```<br>```    return len(images) == len(domain)```<br>```f = lambda x: x * 2```<br>```domain_f = range(1, 10)```<br>```print(is_injective(f, domain_f))``` |
| $f \twoheadrightarrow g$ | Surjective (Onto) Function | Every element of the function's codomain is mapped by at least one element of its domain | ```def is_surjective(func, domain,```<br>```        codomain):```<br>```    images = set(func(x) for x```<br>```        in domain)```<br>```    return images == set(codomain)```<br>```codomain_f = range(2, 20)```<br>```print(is_surjective(f, domain_f,```<br>```        codomain_f))``` |
| $f \rightarrowtail\!\!\!\!\rightarrow g$ or $f \leftrightarrow g$ | Bijective Function | A function that is both one-to-one and onto; i.e., it establishes a perfect pairing between elements of its domain and codomain | ```def is_bijective(func, domain,```<br>```        codomain):```<br>```    return is_injective(func, domain)```<br>```        and is_surjective(func,```<br>```        domain, codomain)```<br>```print(is_bijective(f, domain_f,```<br>```        codomain_f))``` |

## 2.2 Standard Functions

| Symbol | Name | Meaning | Programming Equivalent |
|---|---|---|---|
| $\sqrt{x}$ or $\sqrt[2]{x}$ | Square Root | Principal square root of $x$ | `math.sqrt(x)` |
| $\sqrt[n]{x}$ | n-th Root ($x^{1/n}$) | Principal $n$-th root of $x$ | `x**(1.0/n)` |
| $|x|$ | Absolute Value | Magnitude of a number | `abs(x)` or `numpy.abs(x)` |
| $\sqrt{x}$ | Square Root | Principal square root of $x$ | `math.sqrt(x)` or `numpy.sqrt(x)` |
| $\exp(x)$ | Exponential Function | $e^x$ where $e$ is Euler's number | `math.exp(x)` or `numpy.exp(x)` |
| $\log(x)$ | Natural Logarithm | Logarithm base $e$ | `math.log(x)` or `numpy.log(x)` |
| $\sin(x), \cos(x)$, etc. | Trigonometric Functions | Sine, Cosine, Tangent, Secant, Cosecant, etc. | `math.sin(x)`, `math.cos(x)`, `math.tan(x)`, etc. (or `numpy`) |
| $\sin^{-1}(x)$ or $\arcsin(x)$ | Inverse Trig. Functions | Arcsine, Arccosine, etc. | `numpy.arcsin(x)`, `numpy.arccos(x)`, etc. |
| $\sinh(x), \cosh(x)$, etc. | Hyperbolic Functions | Hyperbolic Sine, Cosine, etc. | `numpy.sinh(x)`, `numpy.cosh(x)`, etc. |
| $\lfloor x \rfloor, \lceil x \rceil, \lfloor x \rceil$ | Floor, Ceiling, and Round | Largest integer $\leq x$ and smallest integer $\geq x$ | `math.floor(x)`, `math.ceil(x)`, `round(x)` |
| $\max(x, y)$ or $\max(S)$ | Maximum | Largest value in a set or between values | `max(x, y)` or `max(S)` |
| $\min(x, y)$ or $\min(S)$ | Minimum | Smallest value in a set or between values | `min(x, y)` or `min(S)` |
| $\text{sgn}(x)$ | Signum | Sign of a number (+ or -) | `numpy.sign(x)` |
| $\text{cis}(x)$ | Cis | Cosine plus $i$ times the sine of $x$ | `numpy.exp(1j * x)` |
| $\text{sinc}(x)$ | Sinc Function | Sine of $x$ divided by $x$ | `numpy.sinc(x)` |
| $\log_a(x)$ | Logarithm Base $a$ | Logarithm of $x$ to the base $a$ | `math.log(x, a)` |
| $\ln(x)$ | Natural Logarithm | Logarithm base $e$ | `math.log(x)` |
| $\text{erf}(x)$ | Error Function | Gaussian error function | `scipy.special.erf(x)` |
| $\Gamma(x)$ | Gamma Function | Generalizes the factorial function for all real numbers | `scipy.special.gamma(x)` |
| $B(x, y)$ | Beta Function | Beta function for $x$ and $y$ | `scipy.special.beta(x, y)` |
| $\zeta(s)$ | Riemann Zeta Function | Same as name | `scipy.special.zeta(s, 1)` |
| $\delta(x)$ | Dirac Delta Function | Delta function, infinite at zero and zero elsewhere | Conceptual in Python; use `scipy.signal.unit_impulse` for discrete approximation |
| $\gcd(a, b)$ | Greatest Common Divisor | Largest positive integer that divides both $a$ and $b$ | `math.gcd(a, b)` |
| $\text{lcm}(a, b)$ | Least Common Multiple | Smallest positive integer that is divisible by both $a$ and $b$ | `numpy.lcm(a, b)` |
| $\phi(n)$ | Euler's Totient Function | Counts the positive integers up to a given integer $n$ that are relatively prime to $n$ | `sympy.totient(n)` |

## 2.3 Combinatorial Functions

| Symbol | Name | Meaning | Programming Equivalent |
|---|---|---|---|
| $S(n)$ | Successor Function | Returns $n+1$ | `S = lambda n: n+1`, then `S(n)` |
| $n!$ | Factorial | The product of all (positive) integers $\leq n$ | `math.factorial(n)` |
| $n!!$ | Double Factorial | The product of all integers of the same parity as $n$ and $\leq$ to $n$ | `scipy.special.factorial2(n)` |
| $\binom{n}{k}$ or $nCk$ or $C(n,k)$ | Binomial Coefficient (Combination) | The number of ways to choose $k$ elements from a set of $n$ elements | `scipy.special.comb(n, k)` |
| $_nP_k$ or $P(n,k)$ | Permutation | The number of ways to arrange $k$ elements out of $n$ | `scipy.special.perm(n, k)` |
| $\binom{n}{k_1,k_2,\ldots,k_m}$ | Multinomial Coefficient | The number of ways to divide a set of $n$ elements into $m$ subsets of sizes $k_1, k_2, \ldots, k_m$ | `K = [k1, k2, ...]`<br>`scipy.special.multinomial(n, K)` |
| $\binom{n+k-1}{k}$ or $\left(\!\binom{n}{k}\!\right)$ | Multichoose | The number of ways to choose $k$ elements from a set of $n$ elements with replacement | `scipy.special.comb(n+k-1, k)` |
| $\binom{n}{k}_q$ | q-binomial Coefficient | A generalization of the binomial coefficient in the context of $q$-calculus | ```def q_binomial(n, k, q):```<br>```    return math.prod(1 - q**(n - i + 1) \```<br>```        for i in range(1, k + 1)) / \```<br>```        math.prod(1 - q**i```<br>```        for i in range(1, k + 1))``` |
| $a \uparrow\uparrow b$ | Tetration (Hyper-4) | Repeated exponentiation of $a$ by itself $b$ times | ```from hyperop import hyperop```<br>```tetration = hyperop(4)```<br>```result = tetration(a, b)``` |
| $!n$ | Derangement | The number of permutations of $n$ elements where no element appears in its original position | ```from scipy.special import perm```<br>```perm(n, n, exact=True) - \```<br>```perm(n, n-1, exact=True)``` |
| $sf(n)$ or $n\$$ | Superfactorial | The product of each number factorial, from 1 to $n$ | ```from math import prod, factorial```<br>```def superfactorial(n):```<br>```    return prod(factorial(i) for i```<br>```        in range(1, n+1))``` |
| $H(n)$ | Hyperfactorial | The product of each number raised to itself, from 1 to $n$ | ```def hyperfactorial(n):```<br>```    return math.prod(i**i for i in```<br>```        range(1, n+1))``` |
| $(n)_k$ | Factorial Power (Falling Factorial) | The product of $k$ consecutive integers starting from $n$ | `scipy.special.poch(n, k)` |
| $n^{(k)}n$ | Rising Factorial (cf. Pochhammer) | Product of $n$ consecutive integers starting from $x$ | `scipy.special.poch(x, n)` |
| $_2F_1(a,b;c;x)$ | Hypergeometric Function | A special function represented by a power series | `scipy.special.hyp2f1(a, b, c, x)` |

## 2.4 Bitwise Functions

| Symbol | Name | Meaning | Programming Equivalent |
|---|---|---|---|
| $a \ \& \ b$ (or) $a$ AND $b$ | Bitwise AND | Same as name | `a & b` |
| $a \mid b$ (or) $a$ OR $b$ | Bitwise OR | Same as name | `a | b` |
| $a \ \hat{} \ b$ (or) $a$ XOR $b$ | Bitwise XOR | Same as name | `a ^ b` |
| $\sim a$ (or) NOT $a$ | Bitwise NOT | One's Complement | `~a` |
| $a \ll b$ (or) $a$ LSH $b$ | Bitwise Shift Left (LSH) | Same as name | `a >> b` |
| $a \gg b$ (or) $a$ RSH $b$ | Bitwise Shift Right (RSH) | Same as name | `a << b` |

## 2.5 Miscellaneous

| Symbol | Name | Meaning | Programming Equivalent |
|---|---|---|---|
| $\arg\min_w f(w)$, $\arg\max_w f(w)$ | Argument of the Minimum/Maximum (For Functions) | The value of the variable ($w$) that minimizes/maximizes a function ($f(w)$) | `from scipy import optimize as o`<br>`# -f(x) for max, f(x) for min`<br>`objective = lambda x: -f(x)`<br>`opt = o.minimize(objective, 0)`<br>`arg_max = opt.x` |
| $\arg\min_i A[i]$, $\arg\max_i A[i]$ | Index of the Minimum/Maximum (For Lists) | The index ($i$) of the minimum/maximum element of a list/array (e.g., $A[i]$; cf. NumPy) | `i = numpy.argmin(A)`,<br>`i = numpy.argmax(A)` |
| $E_k(x)$ | Encryption Function | Encrypts data $x$ using key $k$ | Depends on algorithm; e.g., `rsa.encrypt(msg, pub_key)` |
| $D_k(y)$ | Decryption Function | Decrypts data $y$ using key $k$ | Depends on algorithm; e.g., `rsa.decrypt(cipher, prv_key)` |
| rand() or RNG() | Random Number Generation | Generates a random number; typically in the range [0.0, 1.0] | `random.random()` or `random.randint(a, b)` |
| hash($x$) | Hash Function | Produces a fixed-size hash value from an input $x$ | `hash(x)` or a specific hash function like `hashlib.sha256` |
| char($x$), ord($x$) | Character Encoding | Converts between characters and their numeric codes | `chr(x)`, `ord(x)` |
| $\sum_{n=0}^{\infty} a_n x^n$ | Power Series | Infinite sum of terms in the form of $a_n x^n$ | `def power_series(x, coefficients):`<br>`    return sum(a * (x**n)`<br>`        for n, a in`<br>`        enumerate(coefficients))` |
| $\Delta$ or $D$ | Discriminant | Function of the coefficients of a polynomial, used in formulas to solve algebraic equations | `sympy.discriminant(polynomial)` |
| $\varepsilon$ (or $\varepsilon \to 0$) | Epsilon | Represents an arbitrarily small positive number, often used in limits | `numpy.finfo(float).eps` for machine epsilon |
| $x \gtrless y$ | Greater-Less | Indicates that one quantity is either greater than or less than another | Custom implementation based on context |
| $x \lessgtr y$ | Less-Greater | Indicates that one quantity is either less than or greater than another | Custom implementation based on context |

# 3 Logic

## 3.1 Boolean Operations and Proof Symbols

| Symbol | Name | Meaning | Programming Equivalent |
|---|---|---|---|
| $p \wedge q$ | Logical AND | True if both operands are true | `p and q` |
| $p \vee q$ | Logical OR | True if at least one operand is true | `p or q` |
| $\neg p$ or $\bar{p}$ or $\sim p$ | Logical NOT | True if operand is false | `not p` |
| $p \oplus q$ or $p \veebar q$ | Logical XOR (Exclusive Or) | True if one operand is true but not both | `p ^ q` |
| $p \uparrow q$ or $p \barwedge q$ | Logical NAND (Not And) | True if exactly one operand is false $\neg(p \wedge q)$ | `not (p and q)` |
| $p \downarrow q$ or $p \barvee q$ | Logical NOR | True if both operands are false $\neg(p \vee q)$ | `not (p or q)` |
| $p \equiv q$ or $p \odot q$ | Logical XNOR | True if both operands are the same (true or false) $\neg(p \oplus q)$ | `not (p ^ q)` |
| $p \rightarrow q$ or $p \implies q$ | Implication | True if the first operand implies the second | `p <= q` (or) `(not p) or q` (or equivalent logic) |
| $p \leftarrow q$ or $p \impliedby q$ | Reverse Implication | True if the second operand implies the first | `q <= p` (or) `p or (not q)` (or equivalent logic) |
| $p \iff q$ or $p \leftrightarrow q$ | Biconditional (If and Only If) | True if both operands are the same | `==` (equivalence) |
| $p := q$ or $p \triangleq q$ | Definition | Used to denote (equivalence by) a definition (like "Let p = q" in math) | `p = q` (or) `p := q` (Python 3.8+) |
| $\therefore$ | Therefore | Used to denote a conclusion | `# Therefore` (comment in code) |
| $\because$ | Because | Used to denote a reason | `# Because` (comment in code) |
| ■ or □ | QED | Used to denote the end of a proof | `# QED` (comment in code) |
| $\top$ or $\models S$ | Tautology | Always true | `True` |
| $\bot$ or $\Rightarrow\Leftarrow$ | Contradiction | Always false | `False` |
| $R \vdash S$ | Proves | Assertion in a proof | N/A |
| $R \models S$ | Entails | Logical consequence | N/A |
| $p : q$ or $p \ni q$ | Such That | Logical satisfiability condition | N/A; cf. Set Builder Notation and list comprehensions |

## 3.2 Quantifiers

| Symbol | Name | Meaning | Programming Equivalent |
|---|---|---|---|
| $\forall$ | Universal Quantifier (*For All*) | True if the statement is true for all elements (e.g., $\forall p \in S$) | `all()` in a loop or comprehension (or equivalent `for` loop logic) |
| $\exists$ | Existential Quantifier (*There Exists*) | True if the statement is true for at least one element | `any()` in a loop or comprehension |
| $\nexists$ | Negation of Existential Quantifier | True if no element satisfies the statement | `not any()` or equivalent logic |
| $\exists!$ or $\exists_{=1}$ | Unique Existential Quantifier | True if the statement is true for only one element | `any() and not any()` or equivalent logic |

# 4 Collections

## 4.1 Sets

| Symbol | Name | Meaning | Programming Equivalent |
|---|---|---|---|
| $\emptyset$ or $\{\}$ | Empty Set | A set with no elements | `set(items)` |
| $x \mid p$ or $(x,y) : p$ | Such That (for Sets) | Logical satisfiability condition in Set Builder notation (defines a set of elements using an expression $p$) | Similar to list comprehension; e.g., `S=set([x for x in range(-2, 3)])` is the same as $S = \{x \mid -2 \leq x < 3\}$ |
| $x \in S$ | Element of | Element is a member of (belongs to) a set | `x in S` |
| $x \not\ni S$ or $x \notin S$ | Not Element of | Element is not a member of a set | `x not in S` |
| $A \subseteq B$ | Subset | All elements of the first set are in the second set | `A.issubset(B)` |
| $A \not\subseteq B$ | Not Subset | The first set is not a subset of the second set | `not A.issubset(B)` |
| $A \subset B$ | Proper Subset | All elements of the first set are in the second set and the second set is not exactly equal to the first | `A < B` |
| $A \cup B$ | Union | Set of elements in either or both sets | `A.union(B)` |
| $A \cap V$ | Intersection | Set of elements common to both sets | `A.intersection(B)` |
| $A \setminus B$ or $A - B$ | Set Difference | Elements in the first set but not in the second | `A.difference(B)` |
| $A \supseteq B$ | Superset | All elements of the second set are in the first set | `A.issuperset(B)` |
| $A \supset B$ | Proper Superset | All elements of the second set are in the first set and the first set is not exactly equal to the second | `A > B` |
| $A \uplus B$ | Disjoint Union | Union of two sets that have no elements in common | `A.union(B) if A.isdisjoint(B)` |
| $A \parallel B$ | Disjoint Sets | Two sets with no elements in common | `A.isdisjoint(B)` |
| $\bar{A}$ or $A^c$ | Set Complement | All elements not in the set | `universe.difference(A)` |
| $A \Delta B$ or $A \ominus B$ | Symmetric Difference | Elements in either set, but not in both | `A.symmetric_difference(B)` |
| $A \times B$ | Cartesian Product | Set of all ordered pairs from $A$ and $B$ | `itertools.product(A, B)` |
| $B^A$ | Exponentiation (Set Exp.) | Set of all functions from $A$ to $B$ ($B^A = \{f \mid f : A \to B\}$) | `from itertools import product as itprod`<br>`def set_exponentiation(A, B):  # Concept`<br>`    return set(itprod(B, repeat=len(A)))` |
| $\sup(S)$ | Supremum | Least upper bound of set $S$ | `max(S)` for bounded sets |
| $A \amalg B$ | Amalgamation | Combining multiple sets or structures | `A = {1, 2, 3}`<br>`B = {3, 4, 5}  # Shares a common element`<br>`amalg = A.union(B)  # Conceptually` |
| $\|S\|$ or $\#S$ | Cardinality | The number of elements in a set | `len(S)` |
| $\mathbb{U}$ or $\xi$ or $\mathcal{U}$ | Universal Set | The set of all elements being considered in the current problem | `universe = A.union(B, C, D, ...)` |
| $\mathcal{P}(S)$ or $\mathbb{P}(S)$ | Power Set | Set of all subsets of $S$ | `more_itertools.powerset(S)` |

## 4.2 Lists

| Symbol | Name | Meaning | Programming Equivalent |
|---|---|---|---|
| $(a_1, a_2, \ldots, a_n)$ | List/Tuple | Ordered collection of elements | `list(items)` or `tuple(x, y, ..., z)` |
| $L[i]$ | List Indexing | Accessing the element at index $i$ of list $L$ | `L[i]` |
| $L[i:j]$ | List Slicing | Accessing a subsequence of list $L$ from index $i$ to $j-1$ | `L[i:j]` |
| $L_1 + L_2$ or $L_1 \mid L_2$ | List Concatenation | Combines lists $L_1$ and $L_2$ into a single list | `L1 + L2` |
| $L \times n$ | List Repetition | Repeating list $L$ $n$ times | `L * n` |
| $len(L)$ or $\lvert L \rvert$ | List Length | Number of elements in list $L$ | `len(L)` |
| $L_1 \sqsubseteq L_2$ | Subsequence (s.s. Substring) | List $L_1$ is a subsequence of list $L_2$ (or string is a substring) | `def is_subsequence(L1, L2):`<br>`    iter = iter(L2):`<br>`    return all(x in it for x in L1)`<br>(or)<br>`is_substring = substring in string` |

## 4.3 Big Sums, Products, etc.

| Symbol | Name | Meaning | Programming Equivalent |
|---|---|---|---|
| $\sum_{i=start}^{end} f(x)$ | Summation | Sum over a range of values | `Sum = 0`<br>`for i in range(start, end+1):`<br>`    Sum += expression`<br>(or)<br>`Sum = sum(range(start, end+1))` |
| $\prod_{i=start}^{end} f(x)$ | Product | Product over a range of values | `product = 1`<br>`for i in range(start, end+1):`<br>`    product *= expression`<br>(or)<br>`product = math.prod(range(s, e+1))` |
| $\bigcup_{i=1}^{\infty} A_1$ | Big Union | Union of a set of sets (i.e, $A_1 \cup A_2 \cup A_3 \ldots$) | `def big_union(sets: list):`<br>`    return set.union(*sets)` |
| $\bigcap_{i=1}^{\infty} A_1$ | Big Intersection | Intersection of a set of sets (i.e, $A_1 \cap A_2 \cap A_3 \ldots$) | `def big_intersection(sets: list):`<br>`    return set.intersection(*sets)` |
| $\bigwedge_{i=1}^{n} A_1$ | Big Logical And (or any other operator) | Logical truth of a set of Boolean values (i.e, $p_1 \wedge p_2 \wedge p_3 \wedge \cdots \wedge p_n$) | `from functools import reduce`<br>`from operator import and_, or_`<br>`def big_op(operation, elements):`<br>`    return reduce(operation, elements)`<br>`test_values = [True, True, False]`<br>`big_and = big_op(and_, test_values)`<br>`big_or = big_op(or_, test_values)` |

## 4.4 Graphs and Trees

| Symbol | Name | Meaning | Programming Equivalent |
|---|---|---|---|
| $deg(v)$ | Degree of Vertex | Number of edges incident to vertex $v$ | `len(G.adj[v])` in NetworkX |
| $G(V, E)$ | Graph | Graph $G$ with vertex set $V$ and edge set $E$ | Graph object in a graph library |
| $G_1 \cong G_2$ | Graph Isomorphism | Graph $G_1$ is isomorphic to graph $G_2$ | `networkx.is_isomorphic(G1, G2)` |
| $e = (u, v)$ | Edge | Edge $e$ connecting vertices $u$ and $v$ | Tuple (u, v) in edge list |
| $G[V']$ | Induced Subgraph | Subgraph of $G$ induced by vertex subset $V'$ | `G.subgraph(V')` in NetworkX |
| $d(u, v)$ | Distance | Shortest path distance between vertices $u$ and $v$ | `import networkx as nx` `nx.shortest_path_length(G, u, v)` |
| $\delta(G)$ | Minimum Degree | Smallest degree of any vertex in $G$ | `min(dict(G.degree()).values())` |
| $\Delta(G)$ | Maximum Degree | Largest degree of any vertex in $G$ | `max(dict(G.degree()).values())` |
| $G^c$ | Complement | Graph complement of $G$ | `networkx.complement(G)` |
| $C_n, K_n, P_n$ | Special Graphs | Cycle, Complete, and Path graphs on $n$ vertices | `networkx.cycle_graph(n)`, etc. |
| $T$ | Tree | A connected acyclic graph | Tree data structure |
| $BFS(T)$ | Breadth-First Search | Traverses the tree (or graph) level by level | `networkx.bfs_edges(T, source=root)` |
| $DFS(T)$ | Depth-First Search | Traverses the tree by exploring as far as possible along each branch | `networkx.dfs_edges(T, source=root)` |
| $root(T)$ | Root of Tree | The topmost node in a tree | Access root attribute in tree data structure |
| $children(v)$ | Children of Node | Nodes directly connected to node $v$ downwards | Access children attribute in node data structure |
| $parent(v)$ | Parent of Node | The node directly connected to node $v$ upwards | Access parent attribute in node data structure |
| $depth(v)$ | Depth of Node | Number of edges from $v$ to the tree's root | Recursive function to calculate depth |
| $height(T)$ or $height(v)$ | Height of Tree/Node | Length of the longest downward path to a leaf from node $v$ or the root | Recursive function to calculate height |
| $T_1 \sqcup T_2$ | Tree Union | Union of two trees $T_1$ and $T_2$ | Custom function to merge two trees |
| $LCA(u, v)$ | Lowest Common Ancestor | The lowest node in $T$ that has both $u$ and $v$ as descendants | Custom function or specific tree data structure methods |

# 5 Geometry and Linear Algebra

## 5.1 Vectors

| Symbol | Name | Meaning | Programming Equivalent |
|---|---|---|---|
| $\mathbf{v}$ or $\vec{v}$ | Vector | An ordered tuple of elements | `numpy.array([v1, v2, ..., vn])` |
| $\begin{bmatrix} v_1 \\ v_2 \\ \vdots \\ v_n \end{bmatrix}$ | Column Vector | A vector represented as a column | `numpy.array([[v1], [v2], ..., [vn]])` |
| $[v_1 \; v_2 \; \cdots \; v_n]$ or $\langle v_1, v_2, v_n \rangle$ | Row Vector | A vector represented as a row | `numpy.array([v1, v2, ..., vn])` |
| $\|\mathbf{v}\|$ | Vector Norm | Length or magnitude of the vector | `numpy.linalg.norm(v)` |
| $\mathbf{u} \cdot \mathbf{v}$ | Dot Product | Scalar product of two vectors | `numpy.dot(u, v)` |
| $\mathbf{u} \times \mathbf{v}$ | Cross Product | Vector product of two vectors in 3D space | `numpy.cross(u, v)` |
| $\vec{0}$ or $\mathbf{0}$ | Zero Vector | A vector with all elements equal to zero | `numpy.zeros(n)` |
| $\vec{1}$ or $\mathbf{1}$ | One Vector | A vector with all elements equal to one | `numpy.ones(n)` |
| $\hat{\mathbf{v}}$ | Unit Vector | A vector of length 1 | `v / numpy.linalg.norm(v)` |
| $\mathbf{v}'$ | Translated Vector | A vector that has been moved in space | `v + translation_vector` |
| $\mathbf{v}^*$ | Rotated Vector | A vector that has been rotated | Use rotation matrix or `numpy.dot(rotation_matrix, v)` |
| $\mathbf{u} + \mathbf{v}$ | Vector Sum | Sum of two vectors | `numpy.add(u, v)` |
| $\mathbf{u} - \mathbf{v}$ | Vector Difference | Difference of two vectors | `numpy.subtract(u, v)` |
| $k\mathbf{v}$ | Scalar Multiplication | Multiplying a vector by a scalar ($k$) | `k * v` |
| $\mathbf{v}/k$ or $\frac{v}{k}$ | Scalar Division | Dividing a vector by a scalar | `v / k` |
| $\mathbf{u} \cdot (\mathbf{v} \times \mathbf{w})$ or $[\mathbf{u}, \mathbf{v}, \mathbf{w}]$ | Scalar Triple Product | Volume of the parallelepiped formed by three vectors | `numpy.dot(u, numpy.cross(v, w))` |
| $\|\mathbf{v}\|_p$ | $p$-norm | The $p$-norm of a vector | `numpy.linalg.norm(v, p)` |
| $\mathbf{u} \perp \mathbf{v}$ | Orthogonality | Two vectors are orthogonal | `numpy.dot(u, v) == 0` |
| $\text{proj}_\mathbf{u} \mathbf{v}$ | Projection | Projection of $\mathbf{v}$ onto $\mathbf{u}$ | `numpy.dot(v,u)/numpy.dot(u,u)*u` |
| $\mathbf{u} \prec \mathbf{v}$ or $\mathbf{u} \succ \mathbf{v}$ or $\mathbf{u} \succeq \mathbf{v}$ or $\mathbf{u} \not\preceq \mathbf{v}$, etc. | Vector Ordering | Ordering relations between vectors (Precedes, Succeds) | Custom implementation based on context |
| $\mathcal{V}$ or $V$ | Vector Space | A collection (set) of vectors that can be added together and multiplied by scalars | Conceptual; represented by a set of basis vectors |
| $\mathcal{V}^*$ or $V^*$ | Dual Space | The space of all linear functionals on a vector space | Conceptual; often implemented as functionals in code |
| $S$ | Subspace | A subset of vectors that forms a space | Conceptual; represented by a set of basis vectors |

## 5.2 Matrices

| Symbol | Name | Meaning | Programming Equivalent |
|---|---|---|---|
| $\mathbf{A}$ | Matrix | A rectangular array of numbers | `numpy.array([[a11, a12, ...], ..., [...]])` (or) `np.matrix('a11, a12, a13; a21, ...; ...')` |
| $\mathbf{A}^T$ | Transpose | Flips a matrix over its diagonal | `A.T` (or) `numpy.transpose(A)` |
| $\mathbf{A}^{-1}$ | Matrix Inverse | A matrix that, when multiplied with $\mathbf{A}$, yields the identity | `numpy.linalg.inv(A)` |
| $\mathbf{A} + \mathbf{B}$ | Matrix Addition | Sum of matrices $\mathbf{A}$ and $\mathbf{B}$ | `numpy.add(A, B)` |
| $k\mathbf{A}$ | Scalar Multiplication | Multiplying each element of matrix $\mathbf{A}$ by scalar $k$ | `k * A` |
| $\mathbf{AB}$ or $\mathbf{A} \times \mathbf{B}$ | Matrix Multiplication | Product of matrices $\mathbf{A}$ and $\mathbf{B}$ | `numpy.dot(A, B)` (or) `A @ B` |
| $\mathbf{I}_n$ | Identity Matrix | A square matrix with ones on the diagonal and zeros elsewhere | `numpy.eye(n)` for an $n \times n$ identity matrix |
| $\delta_{ij}$ or $\delta^i_j$ or $\delta(i,j)$ | Kronecker's Delta | Represents the $(i,j)$-entry of an identity matrix | `1 if i == j else 0` |
| $\mathbf{A}^H$ or $\mathbf{A}^*$ | Hermitian | Conjugate transpose of matrix $\mathbf{A}$ | `A.conj().T` |
| $\mathbf{A}^+$ or $\mathbf{A}^\dagger$ | Pseudoinverse | Generalized inverse of matrix $\mathbf{A}$ | `numpy.linalg.pinv(A)` |
| $\mathbf{A} \circ \mathbf{B}$ | Hadamard Product | Element-wise product of $\mathbf{A}$ and $\mathbf{B}$ | `numpy.multiply(A, B)` |
| $\mathbf{J}$ | Matrix of All Ones | A matrix with all elements equal to one | `numpy.ones((m, n))` |
| $\mathbf{A}^n$ | Matrix Powers | Matrix $\mathbf{A}$ raised to the power $n$ | `numpy.linalg.matrix_power(A, n)` |
| $e^{\mathbf{A}}$ | Matrix Exponential | Exponential of matrix $\mathbf{A}$ | `scipy.linalg.expm(A)` |
| $\mathbf{A} \otimes \mathbf{B}$ | Kronecker Product | Tensor product of $\mathbf{A}$ and $\mathbf{B}$ | `numpy.kron(A, B)` |
| $\mathbf{A} \oplus \mathbf{B}$ | Direct Sum | Direct sum of $\mathbf{A}$ and $\mathbf{B}$ | `scipy.linalg.block_diag(A, B)` |
| $\lambda(\mathbf{A})$ | Eigenvalues | Characteristic values of $\mathbf{A}$ | `numpy.linalg.eigvals(A)` |
| $\mathbf{v}(\mathbf{A})$ | Eigenvectors | Characteristic vectors of $\mathbf{A}$ | `numpy.linalg.eig(A)` |
| $\sigma_i(\mathbf{A})$ | Singular Values | Non-negative values that provide insights into the properties of $\mathbf{A}$ | `from numpy.linalg import svd` `svd(A, compute_uv=False)` |
| $\rho(\mathbf{A})$ | Spectral Radius | Largest absolute value of the eigenvalues of $\mathbf{A}$ | `max(abs(numpy.linalg.eigvals(A)))` |
| $\det(\mathbf{A})$ | Determinant | A scalar value representing the volume scaling factor | `numpy.linalg.det(A)` |
| $\dim(\mathbf{A})$ | Dimensionality | Number of rows and columns in $\mathbf{A}$ | `A.shape` |
| $\text{cond}(\mathbf{A})$ | Condition Number | Measure of the sensitivity of the solution of a system of linear equations | `numpy.linalg.cond(A)` |
| $\text{tr}(\mathbf{A})$ | Trace | Sum of the diagonal elements of $\mathbf{A}$ | `numpy.trace(A)` |
| $\text{rank}(\mathbf{A})$ | Rank | Maximum number of linearly independent column vectors in $\mathbf{A}$ | `numpy.linalg.matrix_rank(A)` |
| $\text{null}(\mathbf{A})$ | Nullity | Dimension of the null space of $\mathbf{A}$ | `numpy.linalg.matrix_rank(A)` and subtract from total columns |
| $\ker(\mathbf{A})$ | Kernel | Set of all vectors that map to the zero vector under $\mathbf{A}$ | Use `scipy.linalg.null_space(A)` |
| $\text{perm}(\mathbf{A})$ | Permanent | A function of a square matrix similar to the determinant | `from math import prod` `from itertools import permutations as perm` `def matrix_permanent(A):` `    return sum(prod(A[i, p[i]] for i in range(A.shape[0])) for p in perm(range(A.shape[0])))` |

## 5.3 Matrix Norms

| Symbol | Name | Meaning | Programming Equivalent |
|---|---|---|---|
| $\|\mathbf{A}\|$ or $\|\|\mathbf{A}\|\|$ | Matrix Norm | A measure of the size or length of matrix $\mathbf{A}$ | `numpy.linalg.norm(A)` for Frobenius norm |
| $\|\mathbf{A}\|_1$ | Max Absolute Column Sum Norm | Maximum absolute column sum of the matrix | `numpy.linalg.norm(A, 1)` |
| $\|\mathbf{A}\|_\infty$ | Max Absolute Row Sum Norm | Maximum absolute row sum of the matrix | `numpy.linalg.norm(A, numpy.inf)` |
| $\|\mathbf{A}\|_p$ or $\|\mathbf{A}\|_2$ | Operator Norm (Spectral Norm) | Largest singular value of $\mathbf{A}$ ($p$-norm; 2 for Spectral) | `numpy.linalg.norm(A, 2)` |
| $\|\mathbf{A}\|_F$ | Frobenius Norm | Square root of the sum of the absolute squares of its elements | `numpy.linalg.norm(A, 'fro')` |

## 5.4 Tensors

| Symbol | Name | Meaning | Programming Equivalent |
|---|---|---|---|
| $\mathcal{T}$ | Tensor | A multi-dimensional array of numerical values | `numpy.array([...])` with higher dimensions |
| $\mathcal{T}_1 \otimes \mathcal{T}_2$ | Tensor Product | Product of two tensors, resulting in a new tensor | `numpy.tensordot(T1, T2, axes=0)` |
| $T_{ij}^k$ | Einstein Notation | Notation for writing tensor operations without summation symbols | Implemented using tensor operations in libraries (`numpy`) |

## 5.5 Groups

| Symbol | Name | Meaning | Programming Equivalent |
|---|---|---|---|
| $GL(n, \mathbb{R})$ | General Linear Group | The group of $n \times n$ invertible matrices over the real numbers | Conceptual; represented using matrix libraries |
| $SL(n, \mathbb{R})$ | Special Linear Group | The group of $n \times n$ matrices with determinant 1 over the real numbers | Conceptual; subset of $GL(n, \mathbb{R})$ |
| $O(n)$ | Orthogonal Group | The group of $n \times n$ orthogonal matrices over the real numbers | Conceptual; represented using matrix libraries |
| $SO(n)$ | Special Orthogonal Group | The group of $n \times n$ orthogonal matrices with determinant 1 | Conceptual; subset of $O(n)$ |
| $S_n$ or $\mathfrak{S}_n$ | Symmetric Group | The group of all permutations of $n$ elements | Conceptual; permutations in Python's `itertools` |

## 5.6 Coordinates

| Symbol | Name | Meaning | Programming Equivalent |
|---|---|---|---|
| $(x, y)$ | Cartesian Coordinates | A pair of (or set/vector of 2+) numerical values specifying the position of a point on a plane | `(x, y)` (or) `(x, y, z)` (etc.) |
| $r \angle \theta$ | Polar Coordinates | A system where a point on a plane is determined by a distance from a reference point and an angle from a reference direction | `x = r * math.cos(theta)`<br>`y = r * math.sin(theta)` |
| $(r, \theta, \phi)$ | Spherical Coordinates | A system where a point in 3D space is determined by a distance from a reference point, an angle from a reference direction on the $xy$-plane, and an angle from the $z$-axis | `x = r * math.sin(phi) * math.cos(theta)`<br>`y = r * math.sin(phi) * math.sin(theta)`<br>`z = r * math.cos(phi)` |
| $(r, \theta, z)$ | Cylindrical Coordinates | A system where a point in 3D space is determined by a distance from a reference point on the $xy$-plane, an angle from a reference direction on the $xy$-plane, and a height along the $z$-axis | `x = r * math.cos(theta)`<br>`y = r * math.sin(theta)`<br>`# z remains the same` |

## 5.7 Geometry

| Symbol | Name | Meaning | Programming Equivalent |
|---|---|---|---|
| $|x-y|$ | Distance | The length of the shortest path between two points | `numpy.linalg.norm(x - y)` for Euclidean distance |
| $\mathbb{R}^n$ or $\mathbb{E}^n$ or $E^n$ | Euclidean Plane ($n$-Dimensional) | $n$-dimensional geometric space | Represented by arrays or lists in Python |
| $\overleftrightarrow{AB}$ | Line | Infinite set of points extending in both directions | Represented by a linear equation or function |
| $\overrightarrow{AB}$ | Ray | A line with a starting point extending infinitely in one direction | Represented by a function with a domain constraint |
| $\overset{\frown}{AB}$ | Arc | A segment of a circle's circumference | Represented by a function defining a circular arc |
| $\overline{AB}$ | Line Segment | A part of a line between two endpoints | Represented by two points defining the endpoints |
| $\angle A$ | Angle | The figure formed by two rays sharing a common endpoint | Represented by calculating the angle between vectors |
| $\perp A$ or $\llcorner A$ | Right Angle | An angle of 90 degrees | Check if dot product of direction vectors is 0 |
| $\angle ABC$ or $m\angle ABC$ | Measured Angle | The size of a specific angle, often in degrees or radians | Calculate angle using trigonometric functions |
| $\sphericalangle ABC$ or $\angle_{\text{sph}}ABC$ | Spherical Angle | An angle formed by the intersection of two great circles on a sphere | Calculated using spherical trigonometry |
| $A \cong Y$ | Congruence | Identical in form and size, often used for geometric figures | Check if all corresponding sides and angles are equal |
| $\overline{AB} \perp \overline{CD}$ | Perpendicular | Lines or segments intersecting at a right angle | Check if dot product of direction vectors is 0 |
| $\overline{AB} \parallel \overline{CD}$ | Parallel | Lines or segments that do not intersect | Check if direction vectors are scalar multiples |
| $\triangle ABC$ | Triangle | A polygon with three edges and three vertices | Represented by three points (vertices) |
| $\triangle ABC \sim \triangle DEF$ | Similarity of Triangles | Triangles with the same shape but not exactly the same size | Check if corresponding angles are equal and sides are in proportion |
| $n°$ | Degrees | Unit of measurement for angles (cf. Radians and Gradians) | Use Python's math module for conversion to/from radians |
| $'$ and $''$ | Arc Minute, Arc Second | Units of angular measurement | Used in astronomy; conversion functions required |
| $\mathbf{T}$ | Tangent Vector | A vector that touches a curve at a point and points in the direction of the curve | Calculated using the derivative of the curve's function |
| $\mathbf{N}$ | Normal Vector | A vector perpendicular to the tangent vector at a point on a curve | Calculated as the derivative of the unit tangent vector, normalized |
| $\mathbf{B}$ | Binormal Vector | A vector perpendicular to both the tangent and normal vectors | Calculated as the cross product of the tangent and normal vectors |
| $\mathbb{RP}^2$ | Projective Plane | A geometric surface in which every pair of lines intersects at a point | Conceptual; used in projective geometry |
| $\mathbb{H}$ | Hyperbolic Plane | A surface with constant negative curvature | Represented using hyperbolic geometry models |

# 6 Calculus

## 6.1 Limits

| Symbol | Name | Meaning | Programming Equivalent |
|---|---|---|---|
| $\lim\limits_{x \to a} f(x)$ | Limit | The value that $f(x)$ approaches as $x$ approaches $a$ | `sympy.limit(f, x, a)` |
| $\lim\limits_{x \to \infty} f(x)$ | Limit at Infinity | The value that $f(x)$ approaches as $x$ approaches infinity | `sympy.limit(f, x, sympy.oo)` |
| $\lim\limits_{x \to a^-} f(x)$ | Left-hand Limit | The value that $f(x)$ approaches as $x$ approaches $a$ from the left | `sympy.limit(f, x, a, dir='-')` |
| $\lim\limits_{x \to a^+} f(x)$ | Right-hand Limit | The value that $f(x)$ approaches as $x$ approaches $a$ from the right | `sympy.limit(f, x, a, dir='+')` |
| $\lim\limits_{n \to \infty} a_n$ | Limit of a Sequence | The value that the sequence $a_n$ approaches as $n$ approaches infinity | `sympy.limit_seq(a_n, n)` |
| $\limsup\limits_{n \to \infty} a_n$ | Limit Superior | The supremum of the set of subsequential limits of $a_n$ | `sympy.Limit(a_n, n, sympy.oo, dir='+').doit()` |
| $\liminf\limits_{n \to \infty} a_n$ | Limit Inferior | The infimum of the set of subsequential limits of $a_n$ | `sympy.Limit(a_n, n, sympy.oo, dir='-').doit()` |

## 6.2 Derivatives

### 6.2.1 Single Independent Variable, Scalar- or Vector-Valued

| Symbol | Name | Meaning | Programming Equivalent |
|---|---|---|---|
| $\frac{df}{dx}$ or $f'$ or $\dot{f}$ | Derivative | Rate at which $f$, a function of $x$, changes | `sympy.diff(f, x)` |
| $\frac{d^n f}{dx^n}$ or $f^{(n)}$ | $n$-th Derivative | The $n$-th rate of change of $f$ | `sympy.diff(f, x, n)` |
| $\left.\frac{df}{dx}\right|_{x=a}$ | Derivative at a Point | Rate of change of $f$ at $x = a$ | `sympy.diff(f, x).subs(x, a)` |
| $Df$ or $\nabla f$ | Differential Operator | Operator that maps a function to its derivative | `sympy.Derivative(f, x)` |
| $\frac{d^2 f}{dx^2}$ or $f''$ | Second Derivative | The rate of change of the rate of change of $f$ | `sympy.diff(f, x, 2)` |

### 6.2.2 Multiple Independent Variables, Scalar-Valued

| Symbol | Name | Meaning | Programming Equivalent |
|---|---|---|---|
| $\frac{\partial f}{\partial x}$ or $D_x f$ | Partial Derivative | Rate at which $f$, a function of several variables, changes with respect to one variable | `sympy.diff(f, x)` |
| $\nabla f$ | Gradient | Vector of partial derivatives of $f$ | `sympy.Matrix([f]).jacobian([x, y, ...])` (or) `sympy.gradient(f)` |
| $\frac{\partial^2 f}{\partial x^2}$ | Second Partial Derivative | Second order partial derivative of $f$ with respect to $x$ | `sympy.diff(f, x, 2)` |
| $\nabla^2 f$ | Laplacian | Sum of second partial derivatives of $f$ | `sympy.laplacian(f)` |
| $\text{Hess}(f)$ | Hessian Matrix | Matrix of second-order partial derivatives of $f$ | `sympy.hessian(f, [x, y, ...])` |

## 6.2.3 Multiple Independent Variables, Vector-Valued

| Symbol | Name | Meaning | Programming Equivalent |
|---|---|---|---|
| $\mathbf{f}(\mathbf{x})$ | Vector-Valued Function | A function that maps $\mathbf{x}$ to a vector | Represented by a vector of functions in Python |
| $f_j(\mathbf{x})$ | Component of Vector Function | The $j$-th component of the vector-valued function $\mathbf{f}(\mathbf{x})$ | Each component function in a vector |
| $\frac{\partial \mathbf{f}}{\partial x_i}$ | Partial Derivative of Vector Function | Partial derivative of vector-valued function with respect to variable $x_i$ | `sympy.diff(f_j, x_i)` for each component $f_j$ |
| $D\mathbf{f}$ or $\mathbf{J_f}$ | Jacobian Matrix | Matrix of all first-order partial derivatives of vector-valued function | `sympy.Matrix([f1, f2, ...])`<br>`    .jacobian([x1, x2, ...])` |
| $\nabla \times \mathbf{f}$ | Curl | Measures the rotation of a vector field | `sympy.vector.curl(f)` |
| $\nabla \cdot \mathbf{f}$ | Divergence | Measures the magnitude of a vector field's source or sink at a given point | `sympy.vector.divergence(f)` |
| $\nabla^2 \mathbf{f}$ | Vector Laplacian | Generalization of the Laplacian to vector fields | `sympy.vector.laplacian(f)` |

## 6.3 Integration

| Symbol | Name | Meaning | Programming Equivalent |
|---|---|---|---|
| $\int f(x)\,dx$ | Integral | Area under the curve of $f(x)$ | `sympy.integrate(f, x)` |
| $\int_a^b f(x)\,dx$ | Definite Integral | Area under the curve of $f(x)$ from $a$ to $b$ | `sympy.integrate(f, (x, a, b))` |
| $\iint_D f(x,y)\,dx\,dy$ | Double Integral | Integral of $f(x,y)$ over a domain $D$ in the $xy$-plane | `sympy.integrate(f, (x, x0, x1),`<br>`                (y, y0, y1))` |
| $F(x)\big|_a^b$ | Function Evaluation | Evaluating the antiderivative $F(x)$ at the limits $a$ and $b$ | `F.subs(x, b) - F.subs(x, a)` |
| $\oint_C f(x,y)\,ds$ | Line Integral | Integral of $f(x,y)$ along a curve $C$ | `sympy.integrate(f, (s, s0, s1))` |
| $\oint_S f(x,y,z)\,dS$ | Surface Integral | Integral of $f(x,y,z)$ over surface $S$ | Numerical or specialized symbolic methods |
| $\oiint_S \mathbf{F} \cdot d\mathbf{S}$ | Closed Surface Integral | Integral of a vector field $\mathbf{F}$ over a closed surface $S$ | Numerical or specialized symbolic methods |
| $\iiint_V f(x,y,z)\,dV$ | Volume Integral | Integral of $f(x,y,z)$ over volume $V$ | `sympy.integrate(f, (x, x0, x1),`<br>`                (y, y0, y1), (z, z0, z1))` |

## 6.4 Convolution and Transforms

| Symbol | Name | Meaning | Programming Equivalent |
|---|---|---|---|
| $(f * g)(t)$ | Convolution | Integral expressing the amount of overlap of one function as it is shifted over another | `numpy.convolve(f, g)` |
| $\mathcal{F}\{f(t)\}$ or $\hat{f}$ or $\mathrm{FFT}(f(t))$ | Fourier Transform | Transforms a function of time into a function of frequency | `numpy.fft.fft(f)` |
| $\mathcal{F}^{-1}\{\hat{f}(\omega)\}$ | Inverse FFT | Transforms a function of frequency back into a function of time | `numpy.fft.ifft(f)` |
| $\mathcal{DFT}\{f[n]\}$ | Discrete FFT | Transforms a sequence of values into components of different frequencies | `numpy.fft.fft(f)` |
| $\hat{f}(k)$ | Fourier Coefficients | Coefficients of the Fourier series of a function | `numpy.fft.fft(f)` |
| $\mathcal{L}\{f(t)\}$ | Laplace Transform | Transforms a function of time into a function of complex frequency | `scipy.integrate.laplace(f)` (or) `sympy.laplace_transform(f, t, s)` |
| $\mathcal{H}\{f(t)\}$ | Haar Transform | Transforms a function into a series of square-shaped wavelet functions | `import pywt  # PyWavelets`<br>`coeffs = pywt.wavedec(data, 'haar')` |

# 7 Probability and Statistics

## 7.1 Probability

| Symbol | Name | Meaning | Programming Equivalent |
|---|---|---|---|
| $P(A)$ | Probability of Event | Likelihood of occurrence of event $A$ | `scipy.stats` for distributions |
| $P(A\|B)$ | Conditional Probability | Probability of $A$ given $B$ | Calculated using joint and marginal probabilities |
| $P(A \cap B)$ | Joint Probability | Probability of both $A$ and $B$ occurring | Calculated using multiplication rule |
| $P(A \cup B)$ | Union Probability | Probability of either $A$ or $B$ occurring | Calculated using addition rule |
| $P(A\|B) = \frac{P(B\|A)P(A)}{P(B)}$ | Bayes' Theorem | Relates the probability of $A$ given $B$ with the probability of $B$ given $A$ | Calculated using probabilities or frequency counts |
| $X$ | Random Variable | A variable whose value is subject to variations due to chance | Represented by arrays or functions in Python |
| $\mathbb{E}[X]$ or $E[X]$ | Expected Value | Average value of a random variable $X$ | `numpy.mean(X)` for sample mean |
| $I_A$ | Indicator Random Variable | Variable that is 1 if event $A$ occurs and 0 otherwise | `1 if A else 0` in Python |
| $E[X\|B]$ | Conditional Expectation | Expected value of $X$ given event $B$ | `numpy.mean(X[B])` |
| $\text{Cov}(X,Y)$ | Covariance | Measure of how much two random variables change together | `numpy.cov(X, Y)` |
| $\rho(X,Y)$ or $\text{Corr}(X,Y)$ | Correlation Coefficient | Measure of linear correlation between two variables $X$ and $Y$ | `numpy.corrcoef(X, Y)` |
| $H(X)$ | Entropy | Measure of uncertainty represented by $X$ | `scipy.stats.entropy` |
| $f_X(x)$ | Probability Density Function (PDF) | Function that describes the likelihood of a random variable taking on a value | `scipy.stats` distributions |
| $F_X(x)$ | Cumulative Distribution Function (CDF) | Probability that $X$ will take a value less than or equal to $x$ | `scipy.stats` distributions |
| $p_X(k)$ | Probability Mass Function (PMF) | Probability that a discrete random variable is exactly equal to some value | `scipy.stats` distributions |

## 7.1.1 Distributions

| Symbol | Name | Meaning | Programming Equivalent |
|---|---|---|---|
| $X \sim \mathcal{D}$ | Distribution of X | X follows the distribution $\mathcal{D}$ | `scipy.stats.D` where D is the distribution |
| $X \sim \mathcal{N}(\mu, \sigma^2)$ | Normal Distribution | Distribution defined by mean $\mu$ and variance $\sigma^2$ | `numpy.random.normal(mu, sigma)` |
| $X \sim \mathcal{U}(a, b)$ | Uniform Distribution | Equally likely outcomes between $a$ and $b$ | `numpy.random.uniform(a, b)` |
| $X \sim \Gamma(k, \theta)$ | Gamma Distribution | Generalization of exponential and chi-squared distributions | `numpy.random.gamma(k, theta)` |
| $X \sim \chi^2(k)$ | Chi-Squared Distribution | Distribution of a sum of the squares of k independent standard normal random variables | `numpy.random.chisquare(k)` |
| $X \sim t(\nu)$ | Student's t-Distribution | Distribution arising from estimating the mean of a normally distributed population | `numpy.random.standard_t(nu)` |
| $X \sim F(d_1, d_2)$ | F-Distribution | Distribution arising from comparing two sample variances | `scipy.stats.f(d1, d2)` |
| $X \sim \text{Binomial}(n, p)$ | Binomial Distribution | Number of successes in $n$ trials with success probability $p$ | `numpy.random.binomial(n, p)` |
| $X \sim \text{Poisson}(\lambda)$ | Poisson Distribution | Number of events in fixed interval with mean rate $\lambda$ | `numpy.random.poisson(lambda)` |
| $X \sim \text{Geometric}(p)$ | Geometric Distribution | Number of trials until first success with success probability $p$ | `numpy.random.geometric(p)` |
| $X \sim \text{Hypergeometric}(N, K, n)$ | Hypergeometric Distribution | Number of successes in $n$ draws without replacement from a finite population | `numpy.random.hypergeometric(N, K, n)` |
| $X \sim \text{Bernoulli}(p)$ | Bernoulli Distribution | Distribution of a random variable which takes value 1 with success probability $p$ | `numpy.random.binomial(1, p)` |
| $X_n \xrightarrow{d} X$ | Convergence in Distribution | Sequence of RVs converging in distribution to $X$ | Conceptual; demonstrated through simulations |
| $X_n \xrightarrow{p} X$ | Convergence in Probability | Sequence of RVs converging in probability to $X$ | Conceptual; demonstrated through simulations |
| $X_n \xrightarrow{a.s.} X$ | Almost Sure Convergence | Sequence of RVs converging almost surely to $X$ | Conceptual; demonstrated through simulations |
| $X_n \xrightarrow{L^2} X$ | $L^2$ Convergence | Sequence of RVs converging in the $L^2$ sense to $X$ | Conceptual; demonstrated through simulations |
| IID | Independent and Identically Distributed | Sequence of RVs that are independent and follow the same distribution | Conceptual; used in theoretical explanations and simulations |

## 7.2 Statistics

| Symbol | Name | Meaning | Programming Equivalent |
|---|---|---|---|
| $\mu$ | Population Mean | Average of a sample | `numpy.mean(x)` |
| $\text{Var}(X)$ or $\sigma^2(X)$ | Population Variance | Measure of the spread of a random variable $X$ | `numpy.var(X)` |
| $\sigma(X)$ | Population Standard Deviation | Measure of the amount of variation or dispersion of a set of values | `numpy.std(X)` |
| $\bar{x}$ | Sample Mean | Average of a sample | `numpy.mean(x)` |
| $s^2$ | Sample Variance | Variance of a sample | `numpy.var(x, ddof=1)` |
| $s$ | Sample Standard Deviation | Standard deviation of a sample | `numpy.std(x, ddof=1)` |
| $\tilde{x}$ | Median | Middle value of a sample | `numpy.median(x)` |
| $\hat{x}$ | Mode | Most frequently occurring value in a sample | `scipy.stats.mode(x)` |
| $r$ | Correlation Coefficient | Measure of the linear correlation between two variables | `numpy.corrcoef(x, y)` |
| $t$ | $t$-Statistic | Used in hypothesis testing | `scipy.stats.ttest_ind(x, y)` for independent samples or `scipy.stats.ttest_1samp(x, y)` for one sample |
| $\chi^2$ | Chi-Squared | Used in hypothesis testing for categorical data | `scipy.stats.chi2_contingency(table)` |
| $F$ | $F$-Statistic | Used in analysis of variance (ANOVA) | `scipy.stats.f_oneway(*args)` |
| $Q_1, Q_3$ | Quartiles | Values that divide the data into four equal parts | `numpy.percentile(x, [25, 75])` |
| $z$ | Z-Score | Standard score indicating how many standard deviations an element is from the mean | `(x - numpy.mean(x)) / numpy.std(x)` |
| $H_0$ and $H_a$ or $H_1$ | Hypothesis Testing | Null hypothesis ($H_0$) and alternative hypothesis ($H_a$ or $H_1$) | Conceptual; use statistical tests like `scipy.stats.ttest_ind` |
| $\alpha$ | Type I Error | Probability of rejecting the null hypothesis when it is true | Set as a value in hypothesis testing (commonly 0.05) |
| $\beta$ | Type II Error | Probability of failing to reject the null hypothesis when it is false | Calculated based on power analysis |
| $\epsilon$ | Error or Noise | Random error or noise in data or measurements | Conceptual; often modeled in simulations |
| $\hat{\theta}$ | Estimator | An estimate of a population parameter | Depends on the parameter being estimated; e.g., `numpy.mean(x)` for mean |

# 8 Approximation

## 8.1 Asymptotic Relations

| Symbol | Name | Meaning | Programming Equivalent |
|---|---|---|---|
| $f(x) \sim g(x)$ | Asymptotically Equivalent | $f(x)$ and $g(x)$ are equivalent for large values of $x$ | Conceptual, used in algorithm analysis |
| $f(x) \asymp g(x)$ | Approximate Asymptotic Equality | $f(x)$ is approximately equal to $g(x)$ for large values of $x$ | Conceptual; in programming, compare growth rates or limits |

## 8.2 Big-O Notation and Relatives

| Symbol | Name | Meaning | Programming Equivalent |
|---|---|---|---|
| $O(f(n))$ or $\mathcal{O}(f(n))$ | Big-O Notation | Upper bound on the growth rate of a function (worst case) | Used in algorithm complexity analysis |
| $\Omega(f(n))$ | Big-Omega Notation | Lower bound on the growth rate of a function (best case) | " |
| $\Theta(f(n))$ | Big-Theta Notation | Tight bound on the growth rate of a function (average case) | " |
| $o(f(n))$ | Little-O Notation | Upper bound, not asymptotically tight | " |
| $\omega(f(n))$ | Little-Omega Notation | Lower bound, not asymptotically tight | " |
| $T(n)$ | Recurrence Relation | An equation or inequality that describes a function in terms of its value at smaller inputs | Used in recursive function analysis; e.g., $T(n) = 2T(n/2) + n$ for merge sort |
| $1$ (i.e., $O(1)$) | Constant Complexity | Constant time regardless of input size | Typical for simple operations (e.g., accessing an array element) |
| $n$ | Linear Complexity | Grows linearly with input size | Typical for simple loops |
| $n^2$ | Quadratic Complexity | Grows quadratically with input size | Typical for double loops |
| $n^k$ | Polynomial Complexity | Grows polynomially with input size | Typical for nested loops of depth $k$ |
| $a^n$ | Exponential Complexity | Grows exponentially with input size | Typical for recursive algorithms that double the problem size |
| $\log_b(n)$ | Logarithmic Complexity | Grows logarithmically with input size | Typical for divide-and-conquer algorithms |
| $n \log n$ | Linearithmic Complexity | Grows linearly and logarithmically with input size | Typical for efficient sorting algorithms |
| $\left|\frac{f(x)}{g(x)}\right| \leq b$ | Bounded Ratio | The ratio of $f(x)$ to $g(x)$ is bounded | Conceptual, used in comparing growth rates |
| $\omega(1)$ | Trending to Infinity | A term that grows without bound | Conceptual, indicates unbounded growth |

# 9 Letters

## 9.1 Alphabet Decorators

| Name | Example | Common Uses |
|---|---|---|
| Hat | $\hat{x}$ | Estimators in statistics, unit vectors |
| Bar | $\bar{x}$ | Mean value in statistics |
| Tilde | $\tilde{x}$ | Approximations, equivalences |
| Arrow | $\vec{x}$ | Vectors in physics and engineering |
| Dot | $\dot{x}$ | Derivatives with respect to time |
| Double Dot | $\ddot{x}$ | Second derivatives with respect to time |
| Asterisk | $x^*$ | Complex conjugate, adjoint |
| Prime | $x'$ | Derivatives in mathematics |
| Double Prime | $x''$ | Second derivatives in mathematics |

## 9.2 Latin Alphabet (English)

| Name | Uppercase | Lowercase | Lowercase (Math) | Pronunciation (IPA) |
|---|---|---|---|---|
| A | A | a | $a$ | [eɪ] |
| Bee | B | b | $b$ | [biː] |
| Cee | C | c | $c$ | [siː] |
| Dee | D | d | $d$ | [diː] |
| E | E | e | $e$ | [iː] |
| Eff | F | f | $f$ | [ɛf] |
| Gee | G | g | $g$ | [dʒiː] |
| [H]aitch | H | h | $h$ | [eɪtʃ] or [heɪtʃ] |
| I | I | i | $i$ | [aɪ] |
| Jay | J | j | $j$ | [dʒeɪ] |
| Kay | K | k | $k$ | [keɪ] |
| El | L | l | $l$ | [ɛl] |
| Em | M | m | $m$ | [ɛm] |
| En | N | n | $n$ | [ɛn] |
| O | O | o | $o$ | [oʊ] |
| Pee | P | p | $p$ | [piː] |
| Cue | Q | q | $q$ | [kjuː] |
| Ar | R | r | $r$ | [ɑːr] |
| Ess | S | s | $s$ | [ɛs] |
| Tee | T | t | $t$ | [tiː] |
| U | U | u | $u$ | [juː] |
| Vee | V | v | $v$ | [viː] |
| Double-u | W | w | $w$ | [ˈdʌbəl.juː] |
| Ex | X | x | $x$ | [ɛks] |
| Wye | Y | y | $y$ | [waɪ] |
| Zee/Zed | Z | z | $z$ | [ziː] or [zɛd] |

## 9.3 Greek Alphabet

| Name | Uppercase | Lowercase | Lowercase (Math) | Pronunciation (IPA) |
|---|---|---|---|---|
| Alpha | A | α | $\alpha$ | [ˈælfə] |
| Beta | B | β | $\beta$ | [ˈbeɪːtə] or [ˈbiːtə] |
| Gamma | Γ | γ | $\gamma$ | [ˈgæmə] |
| Delta | Δ | δ | $\delta$ | [ˈdɛltə] |
| Epsilon | E | ϵ | $\epsilon$ or $\varepsilon$ | [ˈɛpsɪlɒn] |
| Zeta | Z | ζ | $\zeta$ | [ˈzeɪːtə] or [ˈziːtə] |
| Eta | H | η | $\eta$ | [ˈeɪːtə] or [ˈiːtə] |
| Theta | Θ | ϑ | $\theta$ or $\vartheta$ | [ˈθeɪːtə] or [ˈθiːtə] |
| Iota | I | ι | $\iota$ | [aɪˈoʊtə] |
| Kappa | K | ϰ | $\kappa$ or $\varkappa$ | [ˈkæpə] |
| Lambda | Λ | λ | $\lambda$ | [ˈlæmdə] |
| Mu | M | μ | $\mu$ | [mjuː] |
| Nu | N | ν | $\nu$ | [njuː] |
| Xi | Ξ | ξ | $\xi$ | [ksi] or [ksaɪ] |
| Omicron | O | o | $o$ | [ˈɒmɪkrɒn] |
| Pi | Π | π | $\pi$ or $\varpi$ | [paɪ] or [pi] |
| Rho | P | ρ | $\rho$ or $\varrho$ | [roʊ] |
| Sigma | Σ | σ/ς | $\sigma$ or $\varsigma$ | [ˈsɪgmə] |
| Tau | T | τ | $\tau$ | [taʊ] or [taf] |
| Upsilon | Υ | υ | $\upsilon$ | [ˈʌpsɪlɒn] |
| Phi | Φ | φ | $\phi$ or $\varphi$ | [fi] or [faɪ] |
| Chi | X | χ | $\xi$ | [xi] or [kaɪ] |
| Psi | Ψ | ψ | $\psi$ | [psi] or [psaɪ] |
| Omega | Ω | ω | $\omega$ | [oʊˈmeɪgə] |

\* For a guide on proper modern Greek pronunciation, see **https://youtu.be/RQF6dZZqX5I**.

# 10 Academic Language

## 10.1 Latin Abbreviations

| Abbreviation | Name | Meaning | Usage |
|---|---|---|---|
| i.e. | id est | "that is" or "in other words" | Used to paraphrase a statement that was just made; **always** followed by a comma (..., *i.e.*, ...) |
| e.g. | exempli gratia | "for example" | Used to give an example of a statement that was just made; **always** followed by a comma |
| viz. | videlicet | "namely" or "more specifically" | Used to clarify a statement that was just made by providing more information; **never** followed by a comma |
| etc. | et cetera | "and so forth" or "and the rest" | Used to suggest that the reader should infer further examples from a list; **usually not** followed by a comma |
| et al. | et alia | "and others" | Used in place of listing multiple authors past the first; **never** followed by a comma |
| cf. | conferre | "see also" or "compare to" or "as seen/found in" | Used to draw a comparison or to refer the reader to an additional source of information; **never** followed by a comma |
| v.s. | vide supra | "see above" | Used to imply that more information can be found before the current point in a written work; **never** followed by a comma |
| q.v. (or 'q.q.' for plural) | quod vide | "which see" or "look it up if interested" | Used to cross-reference a different work or part of a work; **never** followed by a comma |
| n.b. or NB | nota bene | "note well" or "pay attention to the following" | Used to imply that a wise reader will pay especially careful attention to what follows; **never** followed by a comma |
| vs. | versus | "against" or "in contrast to" | Used to contrast two things; **never** followed by a comma |
| c. or ca., cir., or circ. | circa | "around" or "near" | Used when giving an approximation, usually for a date; **never** followed by a comma |
| ex lib. | ex libris | "from the library of" | Used to indicate ownership of a book; **never** followed by a comma |
| s.s. | sensu stricto | "in the strict sense" or "similar/akin to" | Used to specify a narrow, specific definition; **never** followed by a comma |
| s.l. | sensu lato | "in the broad sense" | Used to indicate a broader, more inclusive definition; **never** followed by a comma |
| op. cit. | opere citato | "in the work cited" | Used in bibliographies to refer back to a previously cited source; **never** followed by a comma |
| loc. cit. | loco citato | "in the place cited" | Used to refer to a specific place in a previously cited work; **never** followed by a comma |
| ibid. | ibidem | "in the same place" | Used to refer to the same source as the previous citation (often in a bibliography); **never** followed by a comma |
| id. | idem | "the same" | Used to refer to the same author or work previously mentioned; **never** followed by a comma |
| - | supra | "above" | Used to refer to something mentioned earlier in the text; **never** followed by a comma |
| - | infra | "below" | Used to refer to something mentioned later in the text; **never** followed by a comma |
| - | sic | "so" or "thus" | Used to indicate accuracy in a quoted passage, including errors; **never** followed by a comma |
| Q.E.D. or QED (also ■ or □) | quod erat demonstrandum | "which was to be demonstrated" | Used at the end of a mathematical proof or argument to signify its completion; **never** followed by a comma |

## 10.2 Latin Phrases

| Name | Meaning | Usage |
| --- | --- | --- |
| *a fortiori* | "from the stronger" or "more importantly" | Same as meaning |
| *a priori* | "from before the fact" | Refers to reasoning done before an event happens |
| *a posteriori* | "from after the fact" | Refers to reasoning done after an event |
| *ad hoc* | "for this" | Refers to reasoning that is specific to an event when it is needed and is thus (often) not generalizable to other situations |
| *ad infinitum* | "to infinity" or "without limit" | Same as meaning |
| *ad nauseam* | "to the point of causing nausea" or "to an excessive degree" | Same as meaning |
| *mutatis mutandis* | "changing what needs changing" or "with the necessary changes" | Same as meaning |
| *non sequitur* (sometimes "*non seq.*") | "it does not follow" | Refers to something out of place in a logical argument |
| *de facto* | "in fact" or "in practice" | Refers to something that exists or happens in reality, even if not formally recognized |
| *de jure* | "by law" or "legally" | Refers to something that exists or is such according to law or formal rules |
| *ex ante* | "from before" | Refers to predictions or conditions that exist before a particular event |
| *ex post* | "from after" | Refers to analysis or considerations that occur after an event |
| *in situ* | "in its original place" | Used to describe something that occurs in its original or natural position or place |
| *in vitro* | "in glass" | Refers to processes or reactions occurring in a controlled environment outside of a living organism |
| *in vivo* | "in the living" | Refers to processes or reactions occurring inside a living organism |
| *prima facie* | "at first sight" or "on the face of it" | Refers to something that appears to be true based on first impression or initial evidence |
| *status quo* | "the existing state" | Refers to the existing condition or state of affairs |
| *vice versa* | "the other way around" | Indicates that the relationship or situation is reversible or can be applied in the opposite direction |
| *viva voce* (or "*viva*") | "with living voice" or "by word of mouth" | Refers to an oral examination, especially for a thesis or dissertation defense in academia |
| *Quid quid latine dictum sit, altum videtur* | "Anything said in Latin sounds profound" | Same as meaning |

www.ingramcontent.com/pod-product-compliance
Lightning Source LLC
Chambersburg PA
CBHW081148170526
45158CB00009BA/2767